The ESSENTIALS of

CHEMISTRY

**Staff of Research and Education Association,
Dr. M. Fogiel, Director**

Research and Education Association
505 Eighth Avenue
New York, N.Y. 10018

THE ESSENTIALS OF CHEMISTRY

Printed in the United States of America

Library of Congress Catalog Card Number 87-61814

International Standard Book Number 0-87891-580-X

WHAT "THE ESSENTIALS" WILL DO FOR YOU

This book is a review and study guide. It is comprehensive and it is concise.

It helps in preparing for exams, in doing homework, and remains a handy reference source at all times.

It condenses the vast amount of detail characteristic of the subject matter and summarizes the **essentials** of the field.

It will thus save hours of study and preparation time.

The book provides quick access to the important facts, principles, theorems, concepts, and equations of the field.

Materials needed for exams, can be reviewed in summary form — eliminating the need to read and re-read many pages of textbook and class notes. The summaries will even tend to bring detail to mind that had been previously read or noted.

This "ESSENTIALS" book has been carefully prepared by educators and professionals and was subsequently reviewed by another group of editors to assure accuracy and maximum usefulness.

Dr. Max Fogiel
Program Director

CONTENTS

CHAPTER 1

INTRODUCTION

1.1 MATTER AND ITS PROPERTIES

1.1.1 DEFINITION OF MATTER

Matter occupies space and possesses mass. Mass is an intrinsic property of matter.

Weight is the force, due to gravity, with which an object is attracted to the earth.

Force and mass are related to each other by Newton's equation (Newton's Law), $F = ma$, where F = force, m = mass, and a = acceleration. Weight and mass are related by the equation $w = mg$, where w = weight, m = mass and g = acceleration due to gravity.

Note that the terms "mass" and "weight" are often (incorrectly) used interchangeably through most literature.

1.1.2 STATES OF MATTER

Matter occurs in three states or phases: solid, liquid, and gas. A solid has both a definite size and shape. A liquid has a definite volume but takes the shape of the container, and a gas has neither definite shape nor definite volume.

1.1.3 COMPOSITION OF MATTER

Matter is divided into two categories: distinct substances and mixtures. Distinct substances are either elements or compounds. An element is made up of only one kind of atom. A compound is composed of two or more kinds of atoms joined together in a definite composition.

Mixtures contain two or more distinct substances more or less intimately jumbled together. A mixture has no unique set of properties; it possesses the properties of the substances of which it is composed.

In a homogeneous mixture, the composition and physical properties are ·uniform throughout. Only a single phase is present. A homogeneous mixture can be gaseous, liquid or solid. A heterogeneous mixture, such as oil and water, is not uniform and consists of two or more phases.

1.1.4 PROPERTIES OF MATTER

Extensive properties, such as mass and volume, depend on the size of the sample. Intensive properties, such as melting point, boiling point and density, are independent of sample size.

Physical properties of matter are those properties that can be observed usually with our senses. Examples of physical properties are physical state, color, and melting point.

Chemical properties of a substance are observed only in chemical reactions involving that substance.

Reactivity is a chemical property that refers to the tendency of a substance to undergo a particular chemical reaction.

Chemical changes are those which involve the breaking and/or forming of chemical bonds, as in a chemical reaction.

Physical changes do not result in the formation of new substances. Changes in state are physical changes.

1.2 CONSERVATION OF MATTER

1.2.1 LAW OF CONSERVATION OF MATTER

In a chemical change, matter is neither created nor destroyed, but only changed from one form to another. This law requires that "material balance" be maintained in chemical equations.

1.3 LAWS OF DEFINITE AND MULTIPLE PROPORTIONS

1.3.1 LAW OF DEFINITE PROPORTIONS

A pure compound is always composed of the same elements combined in a definite proportion by mass.

1.3.2 LAW OF MULTIPLE PROPORTIONS

When two elements combine to form more than one compound, different masses of one element combine with a fixed mass of the other element such that those different masses of the first element are in small whole number ratios to each other.

1.4 ENERGY AND CONSERVATION OF ENERGY

1.4.1 DEFINITION OF ENERGY

Energy is usually defined as the ability to do work or transfer heat.

1.4.2 FORMS OF ENERGY

Energy appears in a variety of forms, such as light, sound, heat, mechanical energy, electrical energy and chemical energy. Energy can be converted from one form to another.

Two general classifications of energy are potential energy and kinetic energy. Potential energy is due to position in a field. Kinetic energy is the energy of motion and is equal to one-half of an object's mass, m, multiplied by its speed, s, squared: $KE = \frac{1}{2}ms^2$.

1.4.3 LAW OF CONSERVATION OF ENERGY

Energy can be neither created nor destroyed, but only changed from one form to another.

1.5 MEASUREMENT

Numbers that arise as the result of measurement may contain zero or more significant figures (significant digits). The general rule for multiplication or division is that the product or quotient should not possess any more significant figures than does the least precisely known factor in the calculation.

For addition and subtraction, the rule is that the absolute uncertainty in a sum or difference cannot be smaller than the largest absolute uncertainty in any of the terms in the calculation, i.e., the number of significant figures is limited by the number of digits to the right of the decimal point in the term which is known to the fewest decimal places.

Some examples of assigning significant figures (SF):

Quantity	No. of SF's
.006110	4 6.110×10^{-3}
7,685,000	(4)(?)(ambiguous)
7.685×10^6	4 (ambiguity removed by use of scientific notation)
1.2×10^8	2

4

Examples of arithmetic using the rules for significant figures:

$$(3.39 \times 10^{-3}) \quad (3.0 \times 10^{0}) = 1.1 \times 10^{-3}$$

$$\underbrace{\hspace{2cm}}_{3\ SF} \qquad \underbrace{\hspace{2cm}}_{2\ SF} \qquad \underbrace{\hspace{2cm}}_{2\ SF}$$

$$1.4963 \quad + \quad 0.01 \quad = \quad 1.51$$

4 decimal places   2 decimal places   2 decimal places

1.5.1 LENGTH

1 meter = 39.37 inches, 12 inches = 1 foot,
1 meter = 100 cm, 1 meter = 1000 mm, 1 inch = 2.54 cm

1.5.2 VOLUME

Volume = height × length × width [for a rectangular solid];

πr^{3} [sphere]; $\pi r^{2} \ell$ [right circular cylinder]

1 liter = 1000 cubic centimeters (cc), 1 inch^{3} = 16.4 cc.

1.5.3 MASS

1g = 1000 mg,   1 kg = 1000 g

1.5.4 DENSITY

$$\text{density} = \frac{\text{mass}}{\text{volume}} \left[\frac{g}{ml} \right]$$

1.5.5 TEMPERATURE SCALES

$^{0}C = 5/9(\ ^{0}F - 32\ ^{0})$ Celsius $^{0}F = 9/5(\ ^{0}C)+32\ ^{0}$ Fahrenheit

$^{0}K = \ ^{0}C + 273.15$ Kelvin $^{0}R = \ ^{0}F + 459.67 = 9/5(^{o}K)$ Rankine

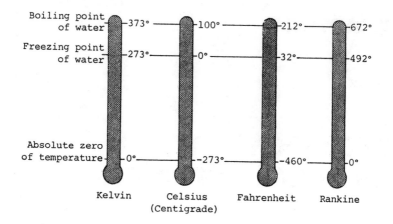

$$3{,}630{,}000 \ = \ 3.630000 \times 10^6, \ .000000123 = 1.23 \times 10^{-7}$$

1.5.7 MULTIPLICATION (EXPONENTS ARE ADDED)

Example: $(5.1 \times 10^{-6})(2 \times 10^{-3}) = 10.2 \times 10^{-9} = 1.02 \times 10^{-8}$

$$= 1 \times 10^{-8} \, (1 \text{ s.f.})$$

1.5.8 DIVISION (EXPONENTS ARE SUBTRACTED)

Example: $(1.5 \times 10^3) \div (5.0 \times 10^{-2})$

$$= .30 \times 10^5 = 3.0 \times 10^4 \ (2 \text{ s.f.})$$

1.5.9 FACTOR LABEL METHOD OF CONVERSION

Dimensional quantities ("units") are treated as algebraic quantities in expressions.

Example:

$$1 \times 10^{-3} \cancel{\text{kilogram}} \times \frac{1 \times 10^3 \ \cancel{\text{grams}}}{1 \ \cancel{\text{kilogram}}} \times \frac{1 \times 10^3 \ \text{milligrams}}{1 \ \cancel{\text{gram}}}$$

$$= 1 \times 10^3 \ \text{mg}$$

CHAPTER 2

STOICHIOMETRY, CHEMICAL ARITHMETIC

2.1 THE MOLE

One mole of any substance is that amount which contains Avogadro's number of particles (atoms, molecules, ions, electrons). Avogadro's number is (approximately) 6.02×10^{23}.

$$\text{number of moles} = \frac{\text{mass}}{\text{molecular weight}}$$

Note the dimensional analysis:

$$\text{mole} = \frac{\text{grams}}{(\text{grams}/\text{mole})}$$

2.2 ATOMIC WEIGHT

The gram-atomic weight of any element is defined as the mass, in grams, which contains one mole of atoms of that element.

For example, approximately 12.0 g of carbon, 16.0 g

7

oxygen, and 32.1 g sulfur, each contain 1 mole of atoms.

(This refers to "average atomic weight"; see also "isotopes".)

2.3 MOLECULAR WEIGHT AND FORMULA WEIGHT

The formula weight (molecular weight) of a molecule or compound is determined by the addition of its component atomic weights.

Example: F.W. of $CaCO_3$ = $1(40)+1(12)+3(16)$ = $100 \frac{g}{mole}$

Molecular weight =

density · volume per molecule · Avogadro's no.

$\underbrace{\qquad\qquad\qquad\qquad\qquad\qquad}$

    mass of one molecule

$\underbrace{\qquad\qquad\qquad\qquad\qquad\qquad\qquad\qquad\qquad\qquad}$

       mass of one mole of molecules

2.4 EQUIVALENT WEIGHT

Equivalent weights are the amounts of substances that react completely with one another in chemical reactions. In electrolysis reactions, the equivalent weight is defined as that weight which either receives or donates 1 mole of electrons (6.022×10^{23} electrons) at an electrode.

For oxidation – reduction reactions, an equivalent is defined as the quantity of a substance that either gains or loses 1 mole of electrons.

In acid-base reactions, an equivalent of an acid is

defined as the quantity of an acid that supplies 1 mole of H^+. An equivalent of a base supplies 1 mole of OH^-.

Note that a given substance may have any of several equivalent weights, depending on the particular reaction in which it is involved. For example, for Fe^{3+}:

$Fe^{3+} + e^- = Fe^{2+}$ one equivalent per mole; eq. wt. = 56 g/eq.

$Fe^{3+} + 3e^- = Fe^0$ three eq. per mole;
$$eq. \ wt. = \left(\frac{1}{3}\right)(56)$$
$$= 18.7 \ g/eq.$$

2.5 BALANCING CHEMICAL EQUATIONS

When balancing chemical equations, one must make sure that there are the same number of atoms of each element on both the left and the right side of the arrow.

Example:

$$2 \ NaOH + H_2SO_4 \rightarrow Na_2SO_4 + 2H_2O.$$

$$\left\{\begin{array}{l} Na: 2 \ atoms \\ O: 6 \ atoms \\ H: 4 \ atoms \\ S: 1 \ atom \end{array}\right\}$$

2.6 CALCULATIONS BASED ON CHEMICAL EQUATIONS

The coefficients in a chemical equation provide the ratio in which moles of one substance react with moles of another.

Example:

$$C_2H_4 + 3O_2 \rightarrow 2CO_2 + 2H_2O \quad \text{represents}$$

$$1 \text{ mole of } C_2H_4 + 3 \text{ moles } O_2 \rightarrow 2 \text{ moles } CO_2$$

$$+ 2 \text{ moles } H_2O.$$

In this equation, the number of moles of O_2 consumed is always equal to three times the number of moles of C_2H_4 that react.

2.7 LIMITING-REACTANT CALCULATIONS

The reactant that is used up first in a chemical reaction is called the limiting reactant, and the amount of product is determined (or limited) by the limiting reactant.

2.8 THEORETICAL YIELD AND PERCENTAGE YIELD

The theoretical yield of a given product is the maximum yield that can be obtained from a given reaction if the reaction goes to completion (rather than to equilibrium).

The percentage yield is a measure of the efficiency of the reaction. It is defined

$$\text{percentage yield} = \frac{\text{actual yield}}{\text{theoretical yield}} \times 100\%$$

2.9 PERCENTAGE COMPOSITION

The percentage composition of a compound is the percentage of the total mass contributed by each element:

$$\% \text{ composition} = \frac{\text{mass of element in compound}}{\text{mass of compound}} \times 100\%$$

2.10 DENSITY AND MOLECULAR WEIGHT

At "STP", standard temperature and pressure, $0\,^0C$ and 760 mm of mercury pressure, 1 mole of any ideal gas occupies 22.4 liters. (The "molar volume" of the gas at STP is 22.4 ℓ/mole.)

The density can be converted to molecular weight using the 22.4 liters/mole relationship:

$$MW = (\text{density})(\text{molar volume})$$

$$\left(\frac{g}{\text{mole}}\right) = \left(\frac{g}{\ell}\right)\left(\frac{\ell}{\text{mole}}\right)$$

2.11 WEIGHT–VOLUME RELATIONSHIPS

For a typical weight-volume problem, follow the following steps:

Step 1: Write the balanced equation for the reaction.

Step 2: Write the given quantities and the unknown quantities for the appropriate substances.

Step 3:   Calculate reacting weights or number of moles (or volume, if the reaction involves only gases) for the substances whose quantities are given. Make sure that the units for each substance are identical.

Step 4:   Use the proportion method or the factor-label method.

Step 5:   Solve for the unknown.

Example: $NaClO_3$, when heated, decomposes to $NaCl$ and O_2. What volume of O_2 at STP results from the decomposition of 42.6 grams $NaClO_3$?

1. (using reactive masses) balanced equation:

$$213g\ NaClO_3 \xrightarrow{\Delta} 117g\ NaCl + 96g\ O_2$$

$$\frac{\text{mass } O_2 \text{ produced}}{\text{mass } NaClO_3 \text{ decomposed}} = \frac{96g}{213g} = \frac{x}{42.6g}$$

$$x = \frac{(96)(42.6)}{213} = 19.2g\ O_2$$

or

2. (using moles) balanced equation:

$$2\ NaClO_3 \xrightarrow{\Delta} 2\ NaCl + 3O_2$$

$$\frac{\text{moles } O_2 \text{ produced}}{\text{moles } NaClO_3 \text{ decomposed}} = \frac{3}{2} = \frac{y}{(42.6g/106.5g/mole)}$$

$$y = \frac{(3)(42.6)}{(2)(106.5)} = 0.6\ \text{mole}\ O_2$$

Finally, at STP, 1 mole (32 g) O_2 occupies 22.4 ℓ, so

$$V_{O_2(STP)} = \left[\frac{19.2g}{32g/mole}\right](22.4\ ℓ/mole) = 0.0268\ ℓ$$

or

$$V_{O_2(STP)} = (0.6\ \text{mole})(22.4\ ℓ/mole) = 0.0268\ ℓ$$

or

$$V_{O_2(STP)} = 2.68 \times 10^{-2}ℓ\ .$$

The ideal gas law, $PV = nRT$, can be used to determine the volume of gas or the number of moles of a gas at conditions other than STP.

CHAPTER 3

ATOMIC STRUCTURE AND THE PERIODIC TABLE

3.1 ATOMIC SPECTRA

The ground state is the lowest energy state available to the atom.

The excited state is any state of energy higher than that of the ground state.

The formula for changes in energy (ΔE) is

$$\Delta E_{electron} = E_{final} - E_{initial}$$

When the electron moves from the ground state to an excited state, it absorbs energy.

When it moves from an excited state to the ground state, it emits energy.

This exchange of energy is the basis for atomic spectra.

3.2 THE BOHR THEORY OF THE HYDROGEN ATOM

Bohr applied to the hydrogen atom the concept that the electron can exist in only certain stable energy levels and that when the electronic state of the atom changes, it must absorb or emit exactly that amount of energy equal to the difference between the final and initial states:

$$\Delta E = E_a - E_b$$

$$E_b - E_a = \frac{z^2 e^2}{2a_0} \left[\frac{1}{n_a{}^2} - \frac{1}{n_b{}^2} \right]$$

measures the energy difference between states a and b, where n = the (quantum) energy level, E = energy, e = charge on electron, a_0 = Bohr radius, and z = atomic number.

The Rydberg-Ritz equation permits calculation of the spectral lines of hydrogen:

$$\frac{1}{\lambda} = R \left[\frac{1}{n_a{}^2} - \frac{1}{n_b{}^2} \right]$$

where R = 109678 cm^{-1} (Rydberg constant), n_a and n_b are the quantum numbers for states a and b, and λ is the wavelength of light emitted or absorbed.

Light behaves as if it were composed of tiny packets, or quanta, of energy (now called "photons").

$$E_{photon} = h\nu$$

where h is Planck's constant and ν is the frequency of light.

$$E = \frac{hc}{\lambda}$$

where c is the speed of light and λ is the wavelength of light.

The electron is restricted to specific energy levels in the atom. Specifically,

$$E = -\frac{A}{n^2}$$

where A is 2.18×10^{-11} erg, and n is the quantum number.

3.3 ELECTRIC NATURE OF ATOMS

3.3.1 BASIC ELECTRON CHARGES

Cathode rays are made up of very small negatively-charged particles named electrons. The cathode is the negative electrode and the anode is the positive electrode.

The nucleus is made up of small positively-charged particles called protons and of neutral particles called neutrons. The proton mass is approximately equal to the mass of the neutron, and is 1837 times the mass of the electron.

3.3.2 COMPONENTS OF ATOMIC STRUCTURE

The number of protons and neutrons in the nucleus is

called the mass number, which corresponds to the isotopic atomic weight. The atomic number is the number of protons found in the nucleus.

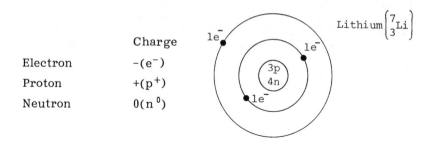

	Charge
Electron	$-(e^-)$
Proton	$+(p^+)$
Neutron	$0(n^0)$

Lithium $\left(^7_3 \text{Li}\right)$

The electrons found in the outermost shell are called valence electrons. When these electrons are lost or partially lost (through sharing), the oxidation state is assigned a positive value for the element. If valence electrons are gained or partially gained by an atom, its oxidation number is taken to be negative.

Example:

$$_{17}\text{Cl} = \bullet)2)8)7 \leftarrow \text{valence electrons.}$$
$$\diagdown\text{nucleus}$$

$(\cdot \overset{\cdot\cdot}{\text{Cl}} :)^-$ This is called the Lewis dot structure of the chloride ion. Its oxidation number is -1.

3.4 THE WAVE MECHANICAL MODEL

Each wave function corresponds to a certain electronic energy and describes a region about the nucleus (called an orbital) where an electron having that energy may be found. The square of the wave function, $|\psi|^2$, is called a probability density, and equals the probability per unit volume of finding the electron in a given region of space.

TABLE 3.1
SUMMARY OF QUANTUM NUMBERS

Principal Quantum Number, n (Shell)	Azimuthal Quantum Number, 1 (Subshell)	Subshell Designation	Magnetic Quantum Number, m (Orbital)	Number of Orbitals in Subshell
1	0	1s	0	1
2	0	2s	0	1
	1	2p	-1,0,+1	3
3	0	3s	0	1
	1	3p	-1,0,+1	3
	2	3d	-2,-1,0,+1,+2	5
4	0	4s	0	1
	1	4p	-1,0,+1	3
	2	4d	-2,-1,0,+1,+2	5
	3	4f	-3,-2,-1,0,+1,+2,+3	7

3.5 SUBSHELLS AND ELECTRON CONFIGURATION

The Pauli exclusion principle states that no two electrons within the same atom may have the same four quantum numbers.

TABLE 3.2 SUBDIVISION OF MAIN ENERGY LEVELS

main energy level	1	2	3	4
number of sublevels(n)	1	2	3	4
number of orbitals(n^2)	1	4	9	16
kind and no. of orbitals per sublevel	s 1	s p 1 3	s p d 1 3 5	s p d f 1 3 5 7
maximum no. of electrons per sublevel	2	2 6	2 6 10	2 6 10 14
maximum no. of electrons per main level ($2n^2$)	2	8	18	32

TABLE 3.3 ELECTRON ARRANGEMENTS

Main Levels	1	2			3	
Sublevels	s	s	p		s	Summary
H	↑					$1s^1$
He	↑↓					$1s^2$
Li	↑↓	↑				$1s^2 2s^1$
Be	↑↓	↑↓				$1s^2 2s^2$
B	↑↓	↑↓	↑ ○ ○			$1s^2 2s^2 2p^1$
C	↑↓	↑↓	↑ ↑ ○			$1s^2 2s^2 2p^2$
N	↑↓	↑↓	↑ ↑ ↑			$1s^2 2s^2 2p^3$
O	↑↓	↑↓	↑↓ ↑ ↑			$1s^2 2s^2 2p^4$
F	↑↓	↑↓	↑↓ ↑↓ ↑			$1s^2 2s^2 2p^5$
Ne	↑↓	↑↓	↑↓ ↑↓ ↑↓			$1s^2 2s^2 2p^6$
Na	↑↓	↑↓	↑↓ ↑↓ ↑↓		↑	$1s^2 2s^2 2p^6 3s^1$
Mg	↑↓	↑↓	↑↓ ↑↓ ↑↓		↑↓	$1s^2 2s^2 2p^6 3s^2$

Hund's Rule states that for a set of equal-energy orbitals, each orbital is occupied by one electron before any orbital has two. Therefore, the first electrons to occupy orbitals within a sublevel have parallel spins. The rule is shown in Table 3.3.

3.6 ISOTOPES

If atoms of the same element (i.e., having identical

atomic numbers) have different masses, they are called isotopes.

The relative abundance of the isotopes is equal to their fraction in the element.

The average atomic weight, A, is equal to

$$M_{avg} = X_1M_1 + X_2M_2 + \ldots + X_NM_N$$

where M_i is the atomic mass of isotope "i" and X_i is the corresponding probability of occurrence.

There are slight differences in chemical behavior of the isotopes of an element. Usually, these differences, called isotope effects, influence the rate of reaction rather than the kind of reaction.

3.7 TRANSITION ELEMENTS AND VARIABLE OXIDATION NUMBERS

Transition elements are elements whose electrons occupy the d sublevel.

Transition elements can exhibit various oxidation numbers. An example of this is manganese, with possible oxidation numbers of +2, +3, +4, +6 and +7.

Groups IB through VIIB and Group VIII constitute the transition elements.

3.8 PERIODIC TABLE

Periodic law states that chemical and physical

properties of the elements are periodic functions of their atomic numbers.

Vertical columns are called groups, each containing a family of elements possessing similar chemical properties.

The horizontal rows in the periodic table are called periods.

The elements lying in two rows just below the main part of the table are called the inner transition elements.

In the first of these rows are elements 58 through 71, called the lanthanides or rare earths.

The second row consists of elements 90 through 103, the actinides.

Group IA elements are called the alkali metals.

Group IIA elements are called the alkaline earth metals.

Group VIIA elements are called the halogens, and the Group O elements, the noble gases.

The metals in the first two groups are the light metals, and those toward the center are the heavy metals. The elements found along the dark line in the chart are called metalloids. They have characteristics of both metals and nonmetals. Some examples of metalloids are boron and silicon.

3.9 PROPERTIES RELATED TO
THE PERIODIC TABLE

The most active metals are found in the lower left corner. The most active nonmetals are found in the upper right corner.

Metallic properties include high electrical conductivity, luster, generally high melting points, ductility (ability to be drawn into wires), and malleability (ability to be hammered into thin sheets). Nonmetals are uniformly very poor conductors of electricity, do not possess the luster of metals and form brittle solids. Metalloids have properties intermediate between those of metals and nonmetals.

3.9.1 ATOMIC RADII

The atomic radius generally decreases across a period from left to right. The atomic radius increases down a group.

3.9.2 ELECTRONEGATIVITY

The electronegativity of an element is a number that measures the relative strength with which the atoms of the element attract valence electrons in a chemical bond. This electronegativity number is based on an arbitrary scale from 0 to 4. Metals have electronegativities less than 2. Electronegativity increases from left to right in a period and decreases as you go down a group.

3.9.3 IONIZATION ENERGY

Ionization energy is defined as the energy required to remove an electron from an isolated atom in its ground state. As we proceed down a group, a decrease in ionization energy occurs. Proceeding across a period from left to right, the ionization energy increases. As we proceed to the right, base-forming properties decrease and acid-forming properties increase.

CHAPTER 4

BONDING

4.1 TYPES OF BONDS

An ionic bond occurs when one or more electrons are transferred from the valence shell of one atom to the valence shell of another.

The atom that loses electrons becomes a positive ion (cation), while the atom that acquires electrons becomes a negatively-charged ion (anion). The ionic bond results from the coulomb attraction between the oppositely-charged ions.

The octet rule states that atoms tend to gain or lose electrons until there are eight electrons in their valence shell.

A covalent bond results from the sharing of a pair of electrons between atoms.

In a nonpolar covalent bond, the electrons are shared equally.

Nonpolar covalent bonds are characteristic of homonuclear diatomic molecules. For example, the flourine molecule:

$$\cdot \ddot{\underset{\cdot\cdot}{F}} : \quad \cdot \ddot{\underset{\cdot\cdot}{F}} : \quad \rightarrow \quad : \ddot{\underset{\cdot\cdot}{F}} : \ddot{\underset{\cdot\cdot}{F}} :$$

Flourine atoms → Flourine molecule

When there is an unequal sharing of electrons between the

atoms involved, the bond is called a polar covalent bond.
An example:

$$H \quad {}^{\times}_{\bullet}\overset{\bullet\bullet}{\underset{\bullet\bullet}{Cl}}{\bullet\bullet} \qquad \begin{array}{l} \times \text{hydrogen electron} \\ \bullet \text{chlorine electrons} \end{array}$$

$$H \quad {}^{\times}_{\bullet}\overset{\bullet\bullet}{\underset{\times\bullet}{O}}{\bullet} \qquad \begin{array}{l} \times \text{hydrogen electron} \\ \bullet \text{oxygen electrons} \end{array}$$
$$\underset{H}{}$$

Because of the unequal sharing, the bonds shown are said to be polar bonds (dipoles). The more electronegative element in the bond is the negative end of the bond dipole. In each of the molecules shown here, there is also a non-zero molecular dipole moment, given by the vector sum of the bond dipoles.

A pure crystal of elemental metal consists of roughly Avogadro's number of atoms held together by metallic bonds.

4.2 INTERMOLECULAR FORCES OF ATTRACTION

A dipole consists of a positive and negative charge separated by a distance. A dipole is described by its dipole moment, which is equal to the charge times the distance between the positive and negative charges:

> net dipole moment = charge × distance

In polar molecular substances, the positive pole of one molecule attracts the negative pole of another. The force of attraction between polar molecules is called a dipolar force.

When a hydrogen atom is bonded to a highly electronegative atom, it will become partially positively-charged, and will be attracted to neighboring electron pairs. This creates a hydrogen bond. The more polar the molecule, the more effective the hydrogen bond is in binding the molecules into a larger unit.

The relatively weak attractive forces between molecules are called Van der Waals forces. These forces become apparent only when the molecules approach one another closely (usually at low temperatures and high pressure). They are due to the way the positive charges of one molecule attract the negative charges of another molecule. Compounds of the solid state that are bound mainly by this type of attraction have soft crystals, are easily deformed, and vaporize easily. Because of the low intermolecular forces, the melting points are low and evaporation takes place so easily that it may occur at room temperature. Examples of substances with this last characteristic are iodine crystals and naphthalene crystals.

4.3 DOUBLE AND TRIPLE BONDS

Sharing two pairs of electrons produces a double bond. An example:

$$O_{\times\times}^{\times\times} : C : {}^{\times\times}_{\times\times}O \qquad \text{or,} \qquad O = C = O$$

The sharing of three electron pairs results in a triple bond. An example:

$$H \overset{\times}{\circ} \ C \overset{\circ}{\circ} \ \overset{\circ}{\circ} C \overset{\times}{\circ} \ H \qquad \text{or,} \qquad H - C \equiv C - H$$

Greater energy is required to break double bonds than single bonds, and triple bonds are harder to break than double bonds. Molecules which contain double and triple bonds have smaller interatomic distances and greater bond strength than molecules with only single bonds. Thus, in the series,

$$H_3C - CH_3, \quad H_2C = CH_2, \quad HC \equiv CH, \qquad \text{the carbon-carbon}$$

distance decreases, and the C-C bond energy increases because of increased bonding.

4.4 RESONANCE STRUCTURES

The resonance structures for sulfur dioxide are as follows:

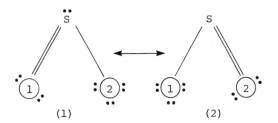

(1)   (2)

The actual electronic structure of SO_2 does not correspond to either 1 or 2 but, instead, to an "average" structure somewhere in between. This true structure is known as a resonance hybrid of the contributing structures 1 and 2.

4.5 ELECTROSTATIC REPULSION AND HYBRIDIZATION

The process of mixing different orbitals of the same atom to form a new set of equivalent orbitals is termed hybridization. The orbitals formed are called hybrid orbitals.

Valence Shell Electron Pair Repulsion (VSEPR) theory permits the geometric arrangement of atoms, or groups of atoms, about some central atom to be determined solely by considering the repulsions between the electron pairs present in the valence shell of the central atom.

Based on VSEPR, the general shape of any molecule can be predicted from the number of bonding and non-bonding electron pairs in the valence shell of the central atom, recalling that nonbonded pairs of electrons (lone pairs) are more repellent than bonded pairs.

TABLE 4.1
SUMMARY OF HYBRIDIZATION

Number of Bonds	Number of Unused e Pairs	Type of Hybrid Orbital	Angle between Bonded Atoms	Geometry	Example
2	0	sp	180^0	Linear	BeF_2
3	0	sp^2	120^0	Trigonal planar	BF_3
4	0	sp^3	109.5^0	Tetrahedral	CH_4
3	1	sp^3	90^0 to 109.5^0	Pyramidal	NH_3
2	2	sp^3	90^0 to 109.5^0	Angular	H_2O
6	0	sp^3d^2	90^0	Octahedral	SF_6

4.6 SIGMA AND PI BONDS

A molecular orbital that is symmetrical around the line passing through two nuclei is called a sigma (σ) orbital. When the electron density in this orbital is concentrated in the bonding region betwen two nuclei, the bond is called a sigma bond.

The bond that is formed by the sideways overlap of two p orbitals, and that provides electron density above and below the line connecting the bound nuclei, is called a π bond (Pi bond).

Pi bonds are present in molecules containing double or triple bonds.

Of the sigma and Pi bonds, the former has greater orbital overlap and is generally the stronger bond.

4.7 PROPERTIES OF IONIC SUBSTANCES

Ionic substances are characterized by the following properties:

1. Ionic crystals have large lattice energies because the electrostatic forces between them are strong.

2. In the solid phase, they are poor electrical conductors.

3. In the liquid phase, they are relatively good conductors of electric current; the mobile charges are the ions (in contrast to metallic conduction, where the electrons constitute the mobile charges).

4. They have relatively high melting and boiling points.

5. They are relatively nonvolatile and have low vapor pressure.

6. They are brittle.

7. Those that are soluble in water form electrolytic solutions that are good conductors of electricity.

4.8 PROPERTIES OF MOLECULAR CRYSTALS AND LIQUIDS

The following are general properties of molecular crystals and/or liquids:

1. Molecular crystals tend to have small lattice energies and are easily deformed because their constituent molecules have relatively weak forces between them.

2. Both the solids and liquids are poor electrical conductors.

3. Many exist as gases at room temperature and atmospheric pressure; those that are solid or liquid at room temperature are relatively volatile.

4. Both the solids and liquids have low melting and boiling points.

5. The solids are generally soft and have a waxy consistency.

6. A large amount of energy is often required to chemically decompose the solids and liquids into simpler substances.

CHAPTER 5

CHEMICAL FORMULAS

5.1 CHEMICAL FORMULAS

A chemical formula is a representation of the make-up of a compound in terms of the kinds of atoms and their relative numbers.

5.2 NAMING COMPOUNDS

Binary compounds consist of two elements. The name shows the two elements present and ends in -ide, such as NaCl = Sodium Chloride. If the metal has only two possible oxidation states, use the suffix -ous for the lower one and -ic for the higher one.

Example: $FeCl_2$ = ferrous chloride [iron(II)chloride]

$FeCl_3$ = ferric chloride [iron(III)chloride]

When naming binary covalent compounds formed between two non-metals, a third system of nomenclature is preferred in which the numbers of each atom in a molecule is specified by a Greek prefix: di(2), tri(3), tetra(4), penta(5) and so on.

Example: N_2O_5 = dinitrogen pentoxide

Ternary compounds, consisting of three elements, are

usually made up of an element and a radical. To name these compounds, you merely name each component, the positive one first and the negative one second.

Binary acids use the prefix hydro- in front of the stem or full name of the non-metallic element, and add the ending -ic.

Example: hydrochloric acid (HCl)

Ternary acids (oxyoacids) usually contain hydrogen, a non-metal, and oxygen. The most common form of the acid consists of merely the stem of the non-metal with the ending -ic. The acid containing one less atom of oxygen than the most common acid has the ending -ous. The acid containing one more atom of oxygen than the most common acid has the prefix per- and the ending -ic. The acid containing one less atom of oxygen than the -ous acid has the prefix hypo- and the ending -ous.

Salts produced by neutralization of acids contain polyatomic anions. The anion derived from the "ic" acid ends in -ate, whereas the anion from the -ous acid ends in -ite

Example: Manganese sulfate ($MnSO_4$)

TABLE 5.1
GROUP VIIA OXY-ACIDS AND OXY-IONS [a]

Oxidation State	Name of Acid		Examples	Name of Anion	
+1	hypo-	-ous	$HClO, HBrO, HIO$	hypo-	-ite
+3		-ous	$HClO_2$		-ite
+5		-ic	$HClO_3, HBrO_3, HIO_3$		-ate
+7	per-	-ic	$HClO_4, HBrO_4, HIO_4,$ H_5IO_6	per-	-ate

[a]The recently discovered HFO is not included because it has yet to be studied in detail.

5.2.1 ACID SALTS

Partial neutralization of an acid that is capable of furnishing more than one H^+ per acid molecule produces salts that are called acid salts.

When only one acid salt is formed, the salt can be named by adding the prefix bi- to the name of the anion of the acid.

Example: $NaHSO_4$ = sodium bisulfate

The salt can also be named by specifying the presence of H, by writing Hydrogen.

Example: Na_2HPO_4 = sodium hydrogen phosphate
(or, disodium hydrogen phosphate)

5.3 WRITING FORMULAS

5.3.1 GENERAL OBSERVATIONS:

1. Metals are assigned positive oxidation numbers while nonmetals (and all the radical ions, except the ammonium ion) are assigned negative oxidation numbers.
2. A radical ion is a group of elements which remain bonded as a group even when involved in formation of compounds.

5.3.2 BASIC RULES FOR WRITING FORMULAS:

1. Represent the symbols of the components using the positive part first, and then the negative part:

   Sodium Chloride, Calcium Oxide, Ammonium Sulfate

   NaCl   CaO $(NH_4)SO_4$

2. Indicate the respective oxidation number, above and to the right of each symbol:

$$Na^{1+} Cl^{1-} , \quad Ca^{2+} O^{2-} , \quad (NH_4)^{1+}(SO_4)^{2-}$$

3. Write the subscript number equal to the oxidation number of the other element or radical:

$$Na^{1+} \cancel{Cl}^{1-}, \qquad Ca^{2+} \underset{\nearrow}{O^{2-}}, \qquad (NH_4)^{1+} \cancel{(SO_4)}^{2-}$$

4. Now rewrite the formulas, omitting the subscript 1, the parentheses of the radicals which have the subscript 1, and the plus and minus signs:

$$NaCl, \quad Ca_2O_2, \quad (NH_4)_2SO_4.$$

5. As a general rule, the subscript numbers in the final formula are reduced to their lowest terms, hence, Ca_2O_2 becomes CaO. There are, however, certain exceptions, such as hydrogen peroxide (H_2O_2) and acetylene (C_2H_2).

5.4 EMPIRICAL AND MOLECULAR FORMULAS

The empirical formula of any compound gives the relative number of atoms of each element in the compound. It is the simplest formula of a material that can be derived solely from its components.

The molecular formula of a substance indicates the actual number of atoms in a molecule of the substance. To determine the molecular formula you must calculate the empirical formula and then extrapolate to the molecular formula via the molecular weight. The molecular formula is a whole number multiple of the empirical formula.

CHAPTER 6

TYPES AND RATES OF CHEMICAL REACTIONS

6.1 TYPES OF CHEMICAL REACTIONS

The four basic kinds of chemical reactions are: combination, decomposition, single replacement, and double replacement. ("Replacement" is sometimes called "metathesis".)

Combination can also be called synthesis. This refers to the formation of a compound from the union of its elements. For example:

$$Zn + S \rightarrow ZnS$$

Decomposition, or analysis, refers to the breakdown of a compound into its individual elements and/or compounds. For example:

$$C_{12}H_{22}O_{11} \rightarrow 12C + 11 H_2O$$

The third type of reaction is called single replacement or single displacement. This type can best be shown by some examples where one substance is displacing another. For example:

$$Fe + CuSO_4 \rightarrow FeSO_4 + Cu$$

The last type of reaction is called double replacement or double displacement, because there is an actual exchange of "partners" to form new compounds. For example:

$$AgNO_3 + NaCl \rightarrow AgCl + NaNO_3$$

6.2 MEASUREMENTS OF REACTION RATES

The measurement of reaction rate is based on the rate of appearance of a product or disappearance of a reactant. It is usually expressed in terms of change in concentration of one of the participants per unit time:

$$\text{rate of chemical reaction} = \frac{\text{change in concentration}}{\text{time}}$$

$$= \frac{\text{moles/liter}}{\text{sec}}$$

For the general reaction $2AB \rightarrow A_2 + B_2$,

$$\text{average rate} = \frac{[AB]_2 - [AB]_1}{t_2 - t_1} = \frac{-\Delta[AB]}{\Delta t}$$

where $[AB]_2$ means "the AB concentration at time t_2".

6.3 FACTORS AFFECTING REACTION RATES

There are five important factors that control the rate of a chemical reaction. These are summarized below:

1. The nature of the reactants and products, i.e., the nature of the transition state formed. Some elements and compounds, because of the bonds broken or formed, react more rapidly with each other than do others.

2. The surface area exposed. Since most reactions depend on the reactants coming into contact, increasing the surface area exposed, proportionally increases the rate of the reaction.

3. The concentrations. The reaction rate usually increases with increasing concentrations of the reactants.

4. The temperature. A temperature increase of $10\,^{0}C$ above room temperature usually causes the reaction rate to double.

5. The catalyst. Catalysts speed up the rate of a reaction but do not change the equilibrium constant (i.e., it simply speeds up the rate of approach to equilibrium).

6.4 THE ARRHENIUS EQUATION: RELATING TEMPERATURE AND REACTION RATE

The following is the Arrhenius equation:

$$k = Ae^{-E_a/RT}$$

where k is the rate constant, A = the Arrhenius constant, E_a = activation energy, R = universal gas constant, and T = temperature in Kelvin. k is small when the activation energy is very large or when the temperature of the reaction mixture is low.

$$\ln k = \ln A - \frac{E_a}{RT}$$
$$y = b + mx$$

A plot of lnk versus 1/T gives a straight line whose slope is equal to $-E_a/R$ and whose intercept with the ordinate is ln A.

$$\ln\left(\frac{k_2}{k_1}\right) = \frac{-E_a}{R}\left[\frac{1}{T_2} - \frac{1}{T_1}\right]$$

$$\log\left(\frac{k_2}{k_1}\right) = \frac{-E_a}{2.303R}\left[\frac{1}{T_2} - \frac{1}{T_1}\right]$$

6.5 ACTIVATION ENERGY

The activation energy is the energy necessary to cause a reaction to occur. It is equal to the difference in energy between the transition state (or "activated complex") and the reactants.

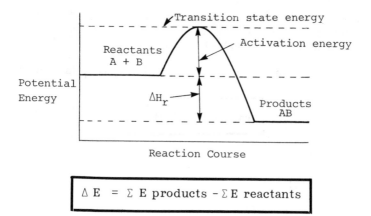

$$\Delta E = \Sigma E \text{ products} - \Sigma E \text{ reactants}$$

In an exothermic process, energy is released and ΔE of reaction is negative; in an endothermic process, energy is absorbed and ΔE is positive.

For a reversible reaction, the energy liberated in the exothermic reaction equals the energy absorbed in the endothermic reaction. (The energy of the reaction, ΔE, is equal also to the difference between the activation energies of the opposing reactions, $\Delta E = E_a - E_a'$).

A catalyst affects a chemical reaction by lowering the activation energy for both the forward and the reverse reactions, equally.

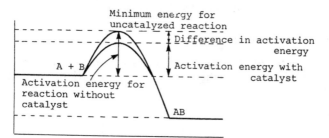

6.6 REACTION RATE LAW

The rate of an irreversible reaction is directly proportional to the concentration of the reactants raised to some power. For the reaction A + B → products, the rate $\propto [A]^x[B]^y$.

The order of the reaction with respect to A is x, and the order with respect to B is y, and the overall order (sum of the individual orders) is x + y.

The following equation is termed the rate law for the reaction:

$$\text{rate} = k[A]^x[B]^y$$

where k is the rate constant.

First order reactions:

$$\text{rate} = k[A]$$

If the reaction rate is doubled by doubling the concentration of the reactant, the order with respect to the reactant is 1.

Second order reactions:

$$\text{rate} = k[A]^2$$

$$\text{rate} = k[2a]^2 = 4 \ ka^2 \quad \text{(Effect of increasing [A] from a to 2a)}$$

If the rate is increased by a factor of four when the concentration of a reactant is doubled, the reaction is second order with respect to that component.

Third order reactions:

The rate of a third order reaction would undergo an eight-fold increase when the concentration is doubled ($2^3 = 8$).

6.7 COLLISION THEORY OF REACTION RATES

The rate of reaction depends on two factors: the number of collisions per unit time, and the fraction of these collisions which result in a reaction.

The following graphs show that the number of collisions, and consequently the rate of reaction, is proportional to the product of the concentrations. The rate of the reaction is directly proportional to the concentration.

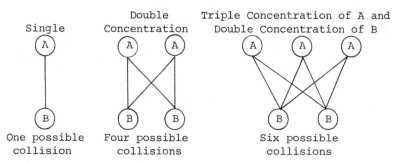

Single	Double Concentration	Triple Concentration of A and Double Concentration of B
One possible collision	Four possible collisions	Six possible collisions

The rate of a reaction according to collision theory is

$$\text{rate} = f \cdot Z$$

where Z is the total number of collisions, and f is the fraction of the total number of collisions occurring at a sufficiently high energy for reaction. The rate of the reaction is further decreased by a factor, p, which is a measure of the importance of the molecular orientations during collision:

$$\text{rate} = pfZ.$$

Z, the collision frequency, is proportional to the concentrations of the reacting molecules:

$$Z = Z_0[A]^n[B]^m$$

where Z_0 is the collision frequency when all of the reactants are at unit concentration.

Therefore, rate = $pfZ_0 [A]^n[B]^m$, or

rate = $k[A]^n[B]^m$, where $k = pfZ_0$.

CHAPTER 7

GASES

7.1 VOLUME AND PRESSURE

A gas has no shape of its own; rather it takes the shape of its container. It has no fixed volume, but is compressed or expanded as its container changes in size. The volume of a gas is the volume of the container in which it is held.

Pressure is defined as force per unit area. Atmospheric pressure is measured using a barometer.

Atmospheric pressure is directly related to the length (h) of the column of mercury in a barometer and is expressed in mm or cm of mercury (hg).

Standard atmospheric pressure is expressed in several ways: 14.7 pounds per square inch (psi), 760mm of mercury, 760 torr or simply 1 "atmosphere" (1 atm).

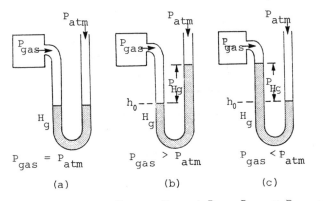

$$P_{gas} = P_{atm} + P_{Hg} \quad P_{gas} = P_{atm} - P_{Hg}$$

Open-end manometer

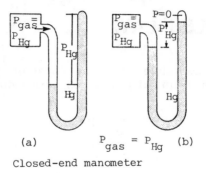

(a) $P_{gas} = P_{Hg}$ (b)

Closed-end manometer

7.2 BOYLE'S LAW

Boyle's law states that, at a constant temperature, the volume of a gas is inversely proportional to the pressure:

$$v \propto \frac{1}{P} \quad \text{or} \quad V = \text{constant} \cdot \frac{1}{P} \quad \text{or} \quad PV = \text{constant}.$$

$$\boxed{P_i V_i = P_f V_f}$$

$$V_f = V_i \left(\frac{P_i}{P_f} \right)$$

A hypothetical gas that would follow Boyle's law under all conditions is called an ideal gas. Deviations from Boyle's law that occur with real gases represent non-ideal behavior.

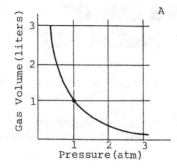

7.3 CHARLES' LAW

Charles' law states that at constant pressure, the volume of a given quantity of a gas varies directly with the temperature :

$V \propto T$ or

$\dfrac{V}{T} = \text{constant.}$

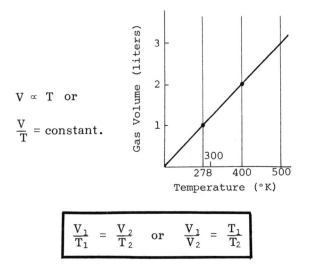

$$\frac{V_1}{T_1} = \frac{V_2}{T_2} \quad \text{or} \quad \frac{V_1}{V_2} = \frac{T_1}{T_2}$$

If Charles' law were strictly obeyed, gases would not condense when they are cooled. This means that gases behave in an ideal fashion only at relatively high temperatures and low pressures.

7.4 DALTON'S LAW OF PARTIAL PRESSURES

The pressure exerted by each gas in a mixture is called its partial pressure. The total pressure exerted by a mixture of gases is equal to the sum of the partial pressures of the gases in the mixture. This statement, known as Dalton's law of partial pressures, can be expressed

$$P_T = Pa + Pb + Pc + \ldots$$

When a gas is collected over water (a typical laboratory method), some water vapor mixes with the gas. The total gas pressure then is given by

$$P_T = P_{gas} + P_{H_2O},$$

where P_{gas} = pressure of dry gas and P_{H_2O} = vapor pressure of water at the temperature of the system.

7.5 LAW OF GAY-LUSSAC

The law of Gay-Lussac states that at constant volume, the pressure exerted by a given mass of gas varies directly with the absolute temperature:

$P \propto T$ (where volume and mass of gas are constant).

$$\frac{P_1}{T_1} = \frac{P_2}{T_2}$$

Gay-Lussac's law of combining volumes states that when reactions take place in the gaseous state, under conditions of constant temperature and pressure, the volumes of reactants and products can be expressed as ratios of small whole numbers.

7.6 IDEAL GAS LAW

$$V \propto \frac{1}{P}, \quad V \propto T, \quad V \propto n$$

then
$$V \propto \frac{nT}{P}$$

$$PV = nRT$$

The hypothetical ideal gas obeys exactly the mathematical statement of the ideal gas law. This statement is also called the equation of state of an ideal gas because it relates the variables (P,V,n,T) that specify properties of the gas. Molecules of ideal gases have no attraction for one another and have no intrinsic volume; they are "point particles". Real gases act in a less than ideal way, especially under conditions of increased pressure and/or decreased temperature. Real gas behavior approaches that of ideal gases as the gas pressure becomes very low. The ideal gas law is thus considered a "limiting law".

7.7 COMBINED GAS LAW

The combined gas law states that for a given mass of gas, the volume is inversely proportional to the pressure and directly proportional to the absolute temperature. This law can be written

$$\frac{P_1V_1}{T_1} = \frac{P_2V_2}{T_2}$$

where P_1 is the original pressure, V_1 is the original volume, T_1 is the original absolute temperature, P_2 is the new pressure, V_2 is the new volume and T_2 is the new absolute temperature.

7.8 AVOGADRO'S LAW (THE MOLE CONCEPT)

Avogadro's law states that under conditions of constant temperature and pressure, equal volumes of different gases contain equal numbers of molecules.

If the initial and final pressure and temperature are the same, then the relationship between the number of molecules, N, and the volume, V, is

$$\frac{V_f}{V_i} = \frac{N_f}{N_i}$$

The laws of Boyle, Charles, Gay-Lussac, and Avogadro are all simple corollaries of the general equation of state for an ideal gas -- PV = nRT -- under various restraining conditions (constant T, constant P, constant V and constant T and P, respectively; n is assumed invariant for all).

7.9 REAL GASES

Real gases fail to obey the ideal gas law under most conditions of temperature and pressure.

Real gases have a finite (non-zero) molecular volume; i.e., they are not true "point particles". The volume within which the molecules may not move is called the excluded volume. The real volume (volume of the container) is therefore slightly larger than the ideal volume (the volume the gas would occupy if the molecules themselves occupied no space):

$$V_{real} = V_{ideal} + nb$$

where b is the excluded volume per mole and n is the number of moles of gas. The ideal pressure, that is, the pressure the gas could exert in the absence of intermolecular attractive forces, is higher than the actual pressure by an amount that is directly proportional to n^2/V^2:

$$P_{ideal} = P_{real} + \frac{n^2 a}{V^2},$$

where a is a proportionality constant that depends on the strength of the intermolecular attractions. Therefore,

$$\left(P + \frac{n^2 a}{V^2} \right) \ (V-nb) = nRT$$

is the Van der Waals equation of state for a real gas.

The values of the constants a and b depend on the particular gas and are tabulated for many real gases.

7.10 GRAHAM'S LAW OF EFFUSION AND DIFFUSION

Effusion is the process in which a gas escapes from one chamber of a vessel to another by passing through a very small opening or orifice.

Graham's law of effusion states that the rate of effusion is inversely proportional to the square root of the density of the gas:

rate of effusion $\propto \sqrt{\dfrac{1}{d}}$, and

$$\frac{\text{rate of effusion (A)}}{\text{rate of effusion (B)}} = \sqrt{\frac{d_B}{d_A}} = \sqrt{\frac{M_B}{M_A}}$$

where M is the molecular weight of each gas, and where the temperature is the same for both gases.

Mixing of molecules of different gases by random motion and collision until the mixture becomes homogeneous is called diffusion.

Graham's law of diffusion states that the relative rates

at which gases will diffuse will be inversely proportional to the square roots of their respective densities or molecular weights:

$$\text{rate} \ \propto \ \frac{1}{\sqrt{\text{mass}}} \quad \text{(where, again, } T_1 = T_2 \text{) and}$$

$$\frac{\text{rate 1}}{\text{rate 2}} = \frac{\sqrt{M_2}}{\sqrt{M_1}} \quad \left(\text{or} \quad \frac{r_1}{r_2} = \frac{\sqrt{d_2}}{\sqrt{d_1}} \right)$$

7.11 THE KINETIC MOLECULAR THEORY

The kinetic molecular theory is summarized as follows:

1. Gases are composed of tiny, invisible molecules that are widely separated from one another in otherwise empty space.

2. The molecules are in constant, continuous, random and straight-line motion.

3. The molecules collide with one another, but the collisions are perfectly elastic (that is, they result in no net loss of energy).

4. The pressure of a gas is the result of collisions between the gas molecules and the walls of the container.

5. The average kinetic energy of all the molecules collectively is directly proportional to the absolute temperature of the gas. The average kinetic energy of equal numbers of molecules of any gas is the same at the same temperature.

CHAPTER 8

LIQUIDS, SOLIDS, AND PHASE CHANGES

8.1 LIQUIDS

A liquid is composed of molecules that are constantly and randomly moving.

8.1.1 VOLUME AND SHAPE

Liquids maintain a definite volume but because of their ability to flow, their shape depends on the contour of the container holding them.

8.1.2 COMPRESSION AND EXPANSION

In a liquid the attractive forces hold the molecules close together, so that increasing the pressure has little effect on the volume. Therefore, liquids are incompressible. Changes in temperature cause only small volume changes.

8.1.3 DIFFUSION

Liquids diffuse much more slowly than gases because of the constant interruptions in the short mean free paths between molecules.

The rates of diffusion in liquids are more rapid at higher temperatures.

8.1.4 SURFACE TENSION

The strength of the inward forces of a liquid is called the liquid's surface tension. Surface tension decreases as the temperature is raised.

8.1.5 KINETICS OF LIQUIDS

Increases in temperature increase the average kinetic energy of molecules and the rapidity of their movement. If a particular molecule gains enough kinetic energy when it is near the surface of a liquid, it can overcome the attractive forces of the liquid phase and escape into the gaseous phase. This is called a change of phase (specifically, evaporation).

8.2 HEAT OF VAPORIZATION AND HEAT OF FUSION

The heat of vaporization of a substance is the number of calories required to convert 1g of liquid to 1g of vapor without a change in temperature.

The reverse process, changing 1g of gas into a liquid without change in temperature, requires the removal of the same amount of heat energy (the heat of condensation).

The heat needed to vaporize 1 mole of a substance is called the molar heat of vaporization or the molar enthalpy of vaporization, ΔH_{vap}, which is also represented as

$$\Delta H_{vaporization} = H_{vapor} - H_{liquid}$$

The magnitude of ΔH_{vap} provides a good measure of the strengths of the attractive forces operative in a liquid.

The number of calories needed to change 1g of a solid substance (at the melting point) to 1g of liquid (at the melting point) is called the heat of fusion.

The total amount of heat that must be removed in order to freeze 1 mole of a liquid is called its molar heat of crystallization. The molar heat of fusion, ΔH_{fus}, is equal in magnitude but opposite in sign to the molar heat of crystallization and is defined as the amount of heat that must be supplied to melt 1 mole of a solid:

$$\Delta H_{fus} = H_{liquid} - H_{solid}$$

8.3 RAOULT'S LAW AND VAPOR PRESSURE

When the rate of evaporation equals the rate of condensation, the system is in equilibrium.

The vapor pressure is the pressure exerted by the gas molecules when they are in equilibrium with the liquid.

The vapor pressure increases with increasing temperature.

Raoult's law states that the vapor pressure of a solution at a particular temperature is equal to the mole fraction of the solvent in the liquid phase multiplied by the vapor pressure of the pure solvent at the same temperature:

$$P_{solution} = X_{solvent} P^0_{solvent}$$

and

$$P_A = X_A P^0_A,$$

where P_A is the vapor pressure of A with solute added, P_A^0 is the vapor pressure of pure A, and X_A is the mole fraction of A in the solution. The solute is assumed here to be nonvolatile (e.g., NaCl or sucrose in water).

8.4 BOILING POINT AND MELTING POINT

The boiling point of a liquid is the temperature at which the pressure of vapor escaping from the liquid equals atmospheric pressure. The normal boiling point of a liquid is the temperature at which its vapor pressure is 760mm Hg, that is, standard atmospheric pressure.

Liquids relatively strong with attractive forces have high boiling points. The melting point of a substance is the temperature at which its solid and liquid phases are in equilibrium.

8.5 SOLIDS

Properties of solids are as follows:

1. They retain their shape and volume when transferred from one container to another.

2. They are virtually incompressible.

3. They exhibit extremely slow rates of diffusion.

In a solid, the attractive forces between the atoms, molecules, or ions are relatively strong. The particles are held in a rigid structural array, wherein they exhibit only vibrational motion.

There are two types of solids, amorphous and crystalline. Crystalline solids are species composed of structural units bounded by specific (regular) geometric patterns. They are characterized by sharp melting points.

Amorphous substances do not display geometric regularity in the solid; glass is an example of an amorphous solid. Amorphous substances have no sharp melting point, but melt over a wide range of temperatures.

When solids are heated at certain pressures, some solids vaporize directly without passing through the liquid phase. This is called sublimation. The heat required to change 1 mole of solid A completely to vapor is called the molar heat of sublimation, ΔH_{sub}. Note that

$$\Delta H_{sub} = \Delta H_{fus} + \Delta H_{vap}$$

8.6 PHASE DIAGRAM

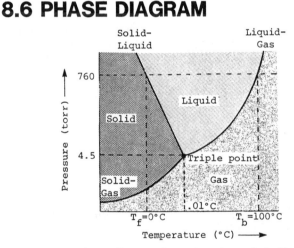

Phase diagram for water (somewhat distorted).

8.7 PHASE EQUILIBRIUM

In a closed system, when the rates of evaporation and condensation are equal, the system is in phase equilibrium.

In a closed system, when opposing changes are taking place at equal rates, the system is said to be in dynamic equilibrium. Virtually all of the equilibria considered in this review are dynamic equilibria.

CHAPTER 9

PROPERTIES OF SOLUTIONS

9.1 TYPES OF SOLUTIONS

There are three types of solutions, gaseous, liquid, and solid.

The most common type of solution consists of a solute dissolved in a liquid.

The atmosphere is an example of a gaseous solution.

Solid solutions, of which many alloys (mixtures of metals) are examples, are of two types:

Substitutional solid solutions in which atoms, molecules, or ions of one substance take the place of particles of another substance in its crystalline lattice.

Interstitial solid solutions are formed by placing atoms of one kind into voids, or interstices, that exist between atoms in the host lattice.

9.2 CONCENTRATION UNITS

Mole fraction is the number of moles of a particular component of a solution divided by the total number of moles of all of the substances present in the solution:

$$X_A = \frac{n_A}{n_A + n_B + n_C + \cdots}$$

$$\sum_{i=1}^{N} X_i = 1$$

Mole percent is equal to 100% × mole fraction. Weight fraction specifies the fraction of the total weight of a solution that is contributed by a particular component. Weight percent is equal to 100% × weight fraction.

Molarity (M) of a solution is the number of moles of solute per liter of solution:

$$\text{Molarity (M)} = \frac{\text{moles of solute}}{\text{liters of solution}}$$

Normality (N) of a solution is the number of equivalents of solute per liter of solution:

$$\text{Normality (N)} = \frac{\text{equiv of solute}}{\text{liters of solution}}$$

Molality of a solution is the number of moles of solute per kilogram (1000g) of solvent.

9.3 THE SOLUTION PROCESS

Solvation is the interaction of solvent molecules with solute molecules or ions to form aggregates, the particles of which are loosely bonded together.

When water is used as the solvent, the process is also called aquation or hydration.

When one substance is soluble in all proportions with

another substance, then the two substances are completely miscible. Ethanol and water are a familiar pair of completely miscible substances.

A saturated solution is one in which solid solute is in equilibrium with dissolved solute.

The solubility of a solute is the concentration of dissolved solute in a saturated solution of that solute.

Unsaturated solutions contain less solute than required for saturation.

Supersaturated solutions contain more solute than required for saturation. Supersaturation is a metastable state; the system will revert spontaneously to a saturated solution (stable state).

9.4 HEATS OF SOLUTION

Heat of solution, ΔH_{soln}, is the quantity of energy that is absorbed or released when a substance enters solution:

$$\Delta H_{soln} = H_{soln} - \Sigma H_{components}$$

The magnitude of the heat of solution provides the information about the relative forces of attraction between the various particles that make up a solution.

Solutions in which the solute–solute, solute–solvent, and solvent–solvent interactions are all the same are called ideal solutions.

9.5 SOLUBILITY AND TEMPERATURE

The solubility of most solids in liquids usually increases with increasing temperature.

For gases in liquids, the solubility usually decreases with increasing temperature.

A positive ΔH^0 indicates that solubility increases with increasing temperature.

$$\log \frac{K_2}{K_1} = \frac{-\Delta H^0}{2.303R} \left[\frac{1}{T_2} - \frac{1}{T_1} \right]$$

Where K_2 = solubility constant at T_2, K_1 = solubility constant at T_1 and ΔH^0 = enthalpy change at standard conditions. For most substances, when a hot concentrated solution is cooled, the excess solid crystallizes. The overall process of dissolving the solute and crystallizing it again is known as recrystallization, and is useful in purification of the solute.

9.6 EFFECT OF PRESSURES ON SOLUBILITY

Pressure has very little effect on the solubility of liquids or solids in liquid solvents.

The solubility of gases in liquid (or solid) solvents always increases with increasing pressure.

9.7 FRACTIONAL CRYSTALLIZATION

The differences in solubility behavior provide the basis for a useful laboratory technique, called fractional crystallization, which is frequently used for the separation of impurities from the products of a chemical reaction.

9.8 FRACTIONAL DISTILLATION

Fractional distillation is used to separate mixtures of volatile liquids into their components.

When a mixture is boiled, it can be separated into two parts: the distillate, which is richer than the original liquid in the more volatile component, and the residue, which is richer in the less volatile component.

9.9 VAPOR PRESSURES OF SOLUTIONS

For a solution in which a nonvolatile solute is dissolved in a solvent, the vapor pressure is due to only the vapor of the solvent above the solution. This vapor pressure is given by Raoult's law:

$$P_{solution} = X_{solvent} \, P^0_{solvent}$$

9.10 COLLIGATIVE PROPERTIES OF SOLUTIONS

Colligative property law: The freezing point, boiling point, and vapor pressure of a solution differ from those of the pure solvent by amounts which are directly proportional to the molal concentration of the solute.

The vapor pressure of an aqueous solution is always lowered by the addition of more solute, which causes the boiling point to be raised (boiling point elevation).

The freezing point is always lowered by addition of

solute (freezing point depression). The freezing point depression, ΔT_f, equals the negative of the molal freezing point depression constant, K_f, times molality(m):

$$\Delta T_f = -K_f(m)$$

The boiling point elevation, ΔT_b, equals the molal boiling point elevation constant, K_b, times molality (m):

$$\Delta T_b = K_b(m)$$

$$\Delta T = T_{solution} - T_{pure\ solvent}$$

9.11 OSMOTIC PRESSURE

Osmosis is the diffusion of a solvent through a semipermeable membrane into a more concentrated solution.

The osmotic pressure of a solution is the minimum pressure that must be applied to the solution to prevent the flow of solvent from pure solvent into the solution.

The osmotic pressure for a solution is:

$$\pi = CRT$$

where π is the osmotic pressure, C is the concentration in molality or molarity, R is the gas constant, and T is the temperature (K). (Note the formal similarity of the osmotic pressure equation to the ideal gas law, $C = n/v$.)

Solutions that have the same osmotic pressure are called isotonic solutions.

Reverse osmosis is a method for recovering pure solvent from a solution.

9.12 INTERIONIC ATTRACTIONS

$$i = \frac{(\Delta T_f) \text{measured}}{(\Delta T_f) \text{calculated as nonelectrolyte}}$$

i, the van't Hoff "factor", is defined as the ratio of the observed freezing point depression produced by a solute in solution, to the freezing point that the solution would exhibit if the solute were a non-electrolyte. For example, since NaCl yields 2 moles of dissolved particles (Na^+ and Cl^-) in water, its van't Hoff factor is 2, and a 1 molal solution of NaCl (aq) yields a freezing point depression which is twice as large as that produced by sucrose, a non-electrolyte.

CHAPTER 10

ACIDS AND BASES

10.1 DEFINITIONS OF ACIDS AND BASES

10.1.1 ARRHENIUS THEORY

The Arrhenius theory states that acids are substances that ionize in water to give H^+ ions, and bases are substances that produce OH^- ions in water.

10.1.2 BRONSTED–LOWRY THEORY

This theory defines acids as proton donors and bases as proton acceptors.

10.1.3 LEWIS THEORY

This theory defines an acid as an electron-pair acceptor and a base as an electron-pair donor.

10.2 PROPERTIES OF ACIDS AND BASES

An acid is a substance which, in aqueous form, conducts electricity, has a sour taste, turns blue litmus red, reacts with active metals to form hydrogen, and neutralizes bases.

A base is a substance which, in aqueous form, conducts electricity, has a bitter taste, turns red litmus blue, feels soapy, and neutralizes acids.

The base that results when an acid donates its proton is called the conjugate base of the acid.

The acid that results when a base accepts a proton is called the conjugate acid of the base.

10.3 FACTORS INFLUENCING THE STRENGTHS OF ACIDS

The greater the number of oxygens bound to the element E in a hydroxy compound, H_xEO_y, the stronger is the acid. This is also a positive correlation with the oxidation state of E.

The acidity of an OH bond in M-O-H depends on the ability of M to draw electrons to itself, thereby weakening the O-H bond, making it more acidic.

With metal ions, the acidity of their solutions depends on the charge on the metal ion.

As we go down a group on the periodic table, the strengths of the oxoacids decrease.

A metal may form acidic hydroxy compounds if the metal has a high oxidation number.

CHAPTER 11

ACID–BASE EQUILIBRIA IN AQUEOUS SOLUTION

11.1 IONIZATION OF WATER , pH

For the equation

$$H_2O + H_2O \rightleftarrows H_3O + OH^-, K_w = [H_3O^+][OH^-] \text{ (or } K_w = [H^+][OH^-])$$

$$= 1.0 \times 10^{-14} \text{at } 25\,^0C,$$

where $[H_3O^+][OH^-]$ is the product of ionic concentrations, and K_w is the ion product constant for water (or, simply the ionization constant or dissociation constant).

$$\boxed{pH = -\log[H^+]}$$

$$\boxed{p\,OH = -\log[OH]}$$

$$\boxed{pK_w = pH + pOH = 14.0}$$

In a neutral solution, pH = 7.0. In an acidic solution pH is less than 7.0. In basic solutions, pH is greater than

7.0. The smaller the pH, the more acidic is the solution. Note that since K_w (like all equilibrium constants) varies with temperature, neutral pH is less than (or greater than) 7.0 when the temperature is higher than (or lower than) $25\,^0C$.

11.2 DISSOCIATION OF WEAK ELECTROLYTES

For the equation

$A^- + H_2O \rightleftarrows HA + OH^-$,

$$K_b = \frac{[HA][OH^-]}{[A^-]}$$

where K_b is the base ionization constant.

$$K_a = \frac{[H^+][A^-]}{[HA]}$$

where K_a is the acid ionization constant.

$$\boxed{K_b = \frac{K_w}{K_a}}$$

for any conjugate acid/base pair, and therefore,

$K_w = [H^+][OH^-]$.

11.3 DISSOCIATION OF POLYPROTIC ACIDS

For $H_2S \rightleftarrows H^+ + HS^-$, $\quad K_{a_1} = \frac{[H^+][HS^-]}{[H_2S]}$

For $HS^- \rightleftharpoons H^+ + S^{2-}$, $\quad K_{a_2} = \dfrac{[H^+][S^{2-}]}{[HS^-]}$

K_{a_1} is much greater than K_{a_2}.

Also, $K_a = K_{a_1} \times K_{a_2} = \dfrac{[H^+][HS^-]}{[H_2S]} \times \dfrac{[H^+][S^{2-}]}{[HS^-]}$

$$= \dfrac{[H^+]^2[S^{2-}]}{[H_2S]}$$

This last equation is useful only in situations where two of the three concentrations are given and you wish to calculate the third.

11.4 BUFFERS

Buffer solutions are equilibrium systems that resist changes in acidity and maintain constant pH when acids or bases are added to them.

The most effective pH range for any buffer is at or near the pH where the acid and salt concentrations are equal (that is, pK_a).

The pH for a buffer is given by

$$pH = pK_a + \log \dfrac{[A^-]}{[HA]} = pK_a + \log \dfrac{[base]}{[acid]}$$

which is obtained very simply from the equation for weak acid equilibrium,

$$K_a = [H+][A^-]/[HA].$$

11.5 HYDROLYSIS

Hydrolysis refers to the action of salts of weak acids or bases with water to form acidic or basic solutions.

Salts of Weak Acids and Strong Bases: Anion Hydrolysis

For $C_2H_3O_2^- + H_2O \rightleftarrows HC_2H_3O_2 + OH^-$,

$$K_h = \frac{[HC_2H_3O_2][OH^-]}{[C_2H_3O_2^-]},$$

where K_h is the hydrolysis constant for the acetate ion, which is just K_b for acetate.

Also,

$$K_a = \frac{[H^+][C_2H_3O_2^-]}{[HC_2H_3O_2]} \quad \text{and}$$

$$\boxed{K_h = \frac{K_w}{K_a}}$$

Salts of Strong Acids and Weak Bases: Cation Hydrolysis

For $NH_4^+ + H_2O \rightleftarrows H_3O^+ + NH_3$,

$$K_h = \frac{[H_3O^+][NH_3]}{[NH_4^+]}$$

($K_h = K_a$ for NH_4^+) Also,

$$\boxed{K_h = \frac{K_w}{K_b}}$$

and

$$K_b = \frac{[NH_4^+][OH^-]}{[NH_3]}$$

Hydrolysis of Salts of Polyprotic Acids

For $S^{2-} + H_2O \; \rightleftarrows \; HS^- + OH^-$,

$$K_{h_1} = \frac{K_w}{K_{a_2}} = \frac{[HS^-][OH^-]}{[S^{2-}]} \, ,$$

where K_{a_2} is the acid dissociation constant for the weak acid HS^-.

For $HS^- + H_2O \; \rightleftarrows \; H_2S + OH^-$,

$$K_{h_2} = \frac{K_w}{K_{a_1}} = \frac{[H_2S][OH^-]}{[HS^-]}$$

where K_{a_1} is the dissociation constant for the weak acid H_2S.

11.6 ACID–BASE TITRATION: THE EQUIVALENT POINT

Titration is the process of determining the amount of a solution of known concentration that is required to react completely with a certain amount of a sample that is being analyzed.

The solution of known concentration is called a standard solution, and the sample being analyzed is the unknown.

$$V_A \times N_A = \text{equiv A, and} \quad V_B \times N_B = \text{equiv B},$$

where V is the volume in liters and N is the normality. An equivalent is defined as the weight in grams of a substance that releases one mole of either protons (H^+) or hydroxyl ions (OH^-).

$$\boxed{V_A \times N_A = V_B \times N_B}$$

at the equivalence point,
since $eq_A = eq_B$

An end point is the point at which a particular indicator changes color.

The equivalent point occurs when equal numbers of equivalents of acid and base have been reacted:

11.6.1 STRONG ACID–STRONG BASE TITRATION

At the equivalent point, the solution is neutral because neither of the ions of the salt in solution undergoes hydrolysis.

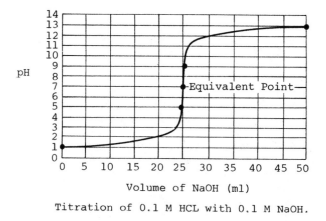

Titration of 0.1 M HCL with 0.1 M NaOH.

11.7 ACID–BASE INDICATORS

Indicators are used to detect the equivalent point in an acid-base titration. They are usually weak organic acids or bases that change color over a narrow pH range ($\Delta pH \sim$ 2pH units).

The choice of a suitable indicator for a particular titration depends on the pH at the equivalence point. The color change of the indicator should occur very near the pH at the equivalence point.

CHAPTER 12

CHEMICAL EQUILIBRIUM

12.1 THE LAW OF MASS ACTION

At equilibrium, both the forward and reverse reactions take place at the same rate, and thus the concentrations of reactants and products no longer change with time.

The law of mass action states that the rate of an elementary chemical reaction is proportional to the product of the concentrations of the reacting substances, each raised to its respective stoichiometric coefficient.

For the reaction $aA + bB \rightleftarrows eE + fF$, at constant temperature,

$$K_c = \frac{[E]^e[F]^f}{[A]^a[B]^b},$$

where the $[...]$ denotes equilibrium molar concentrations, and K_c is the equilibrium constant.

The entire relationship is known as the law of mass action.

$\dfrac{[E]^e[F]^f}{[A]^a[B]^b}$ is known as the mass action expression. Note that if any of the species (A, B, E, F) is a pure solid or pure liquid, it does not appear in the expression for K_c.

For the reaction

$$N_2(g) + 3H_2(g) \rightleftarrows 2NH_3(g),$$

$$K_p = \frac{(P_{NH_3})^2}{P_{N_2}(P_{H_2})^3}$$

where K_p is the equilibrium constant derived from partial pressures.

12.2 KINETICS AND EQUILIBRIUM

The rate of an elementary chemical reaction is proportional to the concentrations of the reactants raised to powers equal to their coeffients in the balanced equation.

For $aA + bB \rightleftarrows eE + fF$,

$$rate_f = k_f[A]^a[B]^b,$$

$$rate_r = k_r[E]^e[F]^f$$

and

$$\boxed{\frac{k_f}{k_r} = \frac{[E]^e[F]^f}{[A]^a[B]^b} = K_c}$$

where K_f and K_r are rate constants for the forward and reverse reactions, respectively.

12.3 THERMODYNAMICS AND CHEMICAL EQUILIBRIUM

$$\boxed{\Delta G = \Delta G^0 + 2.303RT \log Q}$$

The symbol Q represents the mass action expression for the reaction. For gases, Q is written with partial pressures. ΔG is the free energy.

At equilibrium $Q = K_{eq}$ and the products and reactants have the same total free energy, such that $\Delta G = 0$.

$$\Delta G^0 = -2.303RT \log K_{eq} = -RT \ln K_{eq}$$

For the equation $2NO_2(g) \rightleftarrows N_2O_4(g)$,

$$\Delta G^0 = -2.303RT \log \left(\frac{P_{N_2O_4}}{(P_{NO_2})^2} \right)_{eq}, \quad K_c = \frac{[N_2O_4]}{[NO_2]^2}$$

12.4 THE RELATIONSHIP BETWEEN Kp AND Kc

$$K_p = \frac{P_E^e P_F^f}{P_A^a P_B^b} = \frac{[E]^e(RT)^e[F]^f(RT)^f}{[A]^a(RT)^a[B]^b(RT)^b}$$

and

$$K_p = \frac{[E]^e[F]^f}{[A]^a[B]^b} (RT)^{(e+f)-(a+b)}$$

Therefore,

$$\boxed{K_p = K_c(RT)^{\Delta ng}}$$

where Δng is the change in the number of moles of gas upon going from reactants to products.

12.5 HETEROGENEOUS EQUILIBRIA

For heterogeneous reactions, the equilibrium constant expression does not include the concentrations of pure solids and liquids.

For the equation

$$2NaHCO_3(s) \; \rightleftarrows \; Na_2CO_3(s) + CO_2(g) + H_2O(g),$$

$$K_p = P_{CO_2(g)}P_{H_2O(g)},$$

$$K_c = [CO_2(g)][H_2O(g)]$$

and

$$K_p = K_c(RT)^{\Delta ng}, \quad \text{where } \Delta ng = +2 \text{ for the reaction.}$$

12.6 LE CHATELIER'S PRINCIPLE AND CHEMICAL EQUILIBRIUM

Le Chatelier's principle states that when a system at equilibrium is disturbed by the application of a stress (change in temperature, pressure, or concentration) it reacts to minimize the stress and attain a new equilibrium position.

12.6.1 EFFECT OF CHANGING THE CONCENTRATIONS ON EQUILIBRIUM

When a system at equilibrium is disturbed by adding or removing one of the substances, all the concentrations will change until a new equilibrium point is reached with the same value of K_{eq}.

Increase in the concentrations of reactants shifts the equilibrium to the right, thus increasing the amount of products formed. Decreasing the concentrations of reactants shifts the equilibrium to the left and thus decreases the concentrations of products formed.

12.6.2 EFFECT OF TEMPERATURE ON EQUILIBRIUM

An increase in temperature causes the position of equilibrium of an exothermic reaction to be shifted to the left, while that of an endothermic reaction is shifted to the right.

12.6.3 EFFECT OF PRESSURE ON EQUILIBRIUM

Increasing the pressure on a system at equilibrium will cause a shift in the position of equilibrium in the direction of the fewest number of moles of gaseous reactants or products.

12.6.4 EFFECT OF A CATALYST ON THE POSITION OF EQUILIBRIUM

A catalyst lowers the activation energy barrier that must be overcome in order for the reaction to proceed. A catalyst merely speeds the approach to equilibrium, but does not change K_{eq} (or ΔG^0) at all.

12.6.5 ADDITION OF AN INERT GAS

If an inert gas is introduced into a reaction vessel containing other gases at equilibrium, it will cause an increase in the total pressure within the container. However, this kind of pressure increase will not affect the position of equilibrium.

CHAPTER 13

CHEMICAL THERMODYNAMICS

13.1 SOME COMMONLY USED TERMS IN THERMODYNAMICS

A system is that particular portion of the universe on which we wish to focus our attention.

Everything else is called the surroundings.

An adiabatic process occurs when the system is thermally isolated so that no heat enters or leaves.

An isothermal process occurs when the system is maintained at the same temperature throughout an experiment ($t_{final} = t_{initial}$).

An isopiestic (isobaric) process occurs when the system is maintained at constant pressure (i.e., $P_{final} = P_{initial}$).

The state of the system is some particular set of conditions of pressure, temperature, number of moles of each component, and their physical form (for example, gas, liquid, solid or crystalline form).

State functions depend only on the present state of the substance and not on the path by which the present state was attained. Enthalpy, energy, Gibbs free energy, and entropy are examples of state functions.

Heat capacity is the amount of heat energy required to

raise the temperature of a given quantity of a substance one degree celsius.

Specific heat is the amount of heat energy required to raise the temperature of 1g of a substance by $1.0\,^0C$.

Molar heat capacity is the heat necessary to raise the temperature of 1 mole of a substance by $1.0^\circ C$.

13.2 THE FIRST LAW OF THERMODYNAMICS

The first law of thermodynamics states that the change in internal energy is equal to the difference between the energy supplied to the system as heat and the energy removed from the system as work performed on the surroundings:

$$\Delta E = q - w$$

where E represents the internal energy of the system (the total of all the energy possessed by the system). ΔE is the energy difference between the final and initial states of the system:

$$\Delta E = E_{final} - E_{initial}$$

The quantity q represents the amount of heat that is added to the system as it passes from the initial to the final state, and w denotes the work done by the system on the surroundings.

Heat added to a system and work done by a system are considered positive quantities (by convention).

For an ideal gas at constant temperature, $\Delta E = 0$ and $q - w = 0$ ($q = w$).

Considering only work due to expansion of a system, against constant external pressure:

$$w = P_{external} \cdot \Delta V$$

$$\Delta V = V_{final} - V_{initial}$$

13.3 REVERSIBLE AND IRREVERSIBLE PROCESSES

In a reversible expansion of a gas, the opposing pressure is virtually equal to the pressure exerted by the gas. It is reversible because any slight increase in the external pressure will reverse the process and cause compression to occur.

The maximum work derived from any change will be obtained only if the process is carried out in a reversible manner.

$$w_{maximum} \text{ at constant temperature} = 2.30RT \log \frac{V_f}{V_i}$$

All real, spontaneous changes are therefore not reversible, and the work that can be derived from an irreversible change is always less than the theoretical maximum.

13.4 ENTHALPY

The heat content of a substance is called enthalpy, H. A heat change in a chemical reaction is termed a difference in enthalpy, or ΔH. The term "change in enthalpy" refers to the heat change during a process carried out at a constant pressure:

$$\Delta H = q_p \quad , \quad q_p \text{ means "heat at constant pressure"}.$$

The change in enthalpy, ΔH, is defined

$$\Delta H = \Sigma H_{products} - \Sigma H_{reactants}$$

When more than one mole of a compound is reacted or formed, the molar enthalpy of the compound is multiplied by the number of moles reacted (or formed).

Enthalpy is a state function. Changes in enthalpy for exothermic and endothermic reactions are shown below:

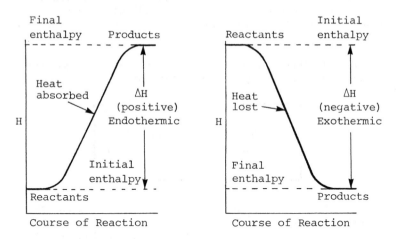

The ΔH of an endothermic reaction is positive, while that for an exothermic reaction is negative.

13.5 HEATS OF REACTION

ΔE is equal to the heat absorbed or evolved by the system under conditions of constant volume:

$\Delta E = q_v$, q_v means "heat at constant volume".

Since

$$H = E + PV$$

at constant pressure $\Delta H = \Delta E + P \Delta V$. Note that the term $P \Delta V$ is just the pressure-volume work $((\Delta n)RT)$ for an ideal gas at constant temperature , where Δn is the number of moles of gaseous products minus the number of moles of gaseous reactants. Therefore,

$$\Delta H = \Delta E + \Delta nRT$$

for a reaction which involves gases. If only solid and liquid phases are present, ΔV is very small, so that $\Delta H \approx \Delta E$.

13.6 HESS' LAW OF HEAT SUMMATION

Hess's law of heat summation states that when a reaction can be expressed as the algebraic sum of two or more reactions, the heat of the reaction, ΔH_r, is the algebraic sum of the heats of the constituent reactions.

The enthalpy changes associated with the reactions that correspond to the formation of a substance from its free elements are called heats of formation, ΔH_f.

$$\Delta H_r^0 = \Sigma \Delta H_f^0 \text{ (products)} - \Sigma \Delta H_f^0 \text{ (reactants)}$$

13.7 STANDARD STATES

The standard state corresponds to $25\,^0C$ and 1 atm.

Heats of formation of substances in their standard states are indicated as ΔH_f^0.

13.8 BOND ENERGIES

For diatomic molecules, the bond dissociation energy, ΔH_{diss}^0, is the amount of energy per mole required to break the bond and produce two atoms, with reactants and products being ideal gases in their standard states at $25\,^0C$.

The heat of formation of an atom is defined as the amount of energy required to form 1 mole of gaseous atoms from its element in its common physical state at $25\,^0C$ and 1 atm pressure.

In the case of diatomic gaseous molecules of elements, the ΔH_f^0 of an atom is equal to one-half the value of the dissociation energy, that is,

$$-\Delta H_f^0 = \tfrac{1}{2} \Delta H_{diss}^0$$

where the minus sign is needed since one process liberates heat, and the other requires heat.

For the reaction $HO\text{-}OH(g) \rightarrow 2OH(g)$,

$$\Delta H_{diss}^0 = 2\left(\Delta H_f^0 OH\right) - \left(\Delta H_f^0\ OH\text{-}OH\right).$$

The energy needed to reduce a gaseous molecule to neutral gaseous atoms, called the atomization energy, is the sum of all of the bond energies in the molecule.

For polyatomic molecules, the average bond energy, ΔH_{diss}^0 avg., is the average energy per bond required to dissociate 1 mole of molecules into their constituent atoms.

13.9 SPONTANEITY OF CHEMICAL REACTIONS

In any spontaneous change, the amount of free energy available decreases toward zero as the process proceeds towards equilibrium.

A negative value of ΔG indicates that a reaction can take place spontaneously, but it does not guarantee that the reaction will take place at all.

13.10 ENTROPY

The degree of randomness of a system is represented by a thermodynamic quantity called the entropy, S. The greater the randomness, the greater the entropy.

A change in entropy or disorder associated with a given system is

$$\Delta S = S_2 - S_1$$

The entropy of the universe increases for any spontaneous process:

$$\Delta S_{universe} = (\Delta S_{system} + \Delta S_{surroundings}) \geq 0$$

When a process occurs reversibly at constant temperature, the change in entropy, ΔS, is equal to the heat absorbed divided by the absolute temperature at which the change occurs:

$$\Delta S = \frac{q_{reversible}}{T}$$

13.11 THE SECOND LAW OF THERMODYNAMICS

The second law of thermodynamics states that in any spontaneous process there is an increase in the entropy of the universe ($\Delta S_{total} > 0$).

$$\Delta S_{universe} = \Delta S_{total} = \Delta S_{system} + \Delta S_{surroundings}$$

$$\Delta S_{surroundings} = \frac{-\Delta H_{system}}{T} \quad \text{at constant P and T}$$

$$T \Delta S_{total} = -(\Delta H_{system} - T \Delta S_{system})$$

The maximum amount of useful work that can be done by any process at constant temperature and pressure is called the change in Gibbs free energy, ΔG;

$$\boxed{\Delta G = \Delta H - T \Delta S}$$

Another way in which the second law is stated is that in any spontaneous change, the amount of free energy available decreases.

Thus, if $\Delta G = 0$, then the system is at equilibrium.

13.12 STANDARD ENTROPIES AND FREE ENERGIES

The entropy of a substance, compared to its entropy in a perfectly crystalline form at absolute zero, is called its absolute entropy, S^0.

The third law of thermodynamics states that the entropy of any pure, perfect crystal at absolute zero is equal to zero.

The standard free energy of formation, ΔG_f^0, of a substance is defined as the change in free energy for the reaction in which one mole of a compound is formed from its elements under standard conditions:

$$\Delta G_f^0 = \Delta H_f^0 - T \Delta S_f^0$$

$$\Delta S_f^0 = \Sigma S_f^0 \text{ products} - \Sigma S_f^0 \text{ reactants}$$

and

$$\Delta G_r^0 = \Sigma \Delta G_f^0 \text{ products} - \Sigma \Delta G_f^0 \text{ reactants}$$

CHAPTER 14

OXIDATION AND REDUCTION

14.1 OXIDATION AND REDUCTION

Oxidation is defined as a reaction in which atoms or ions undergo an increase in oxidation state. The agent that causes oxidation to occur is called the oxidizing agent and is itself reduced in the process.

Reduction is defined as a reaction in which atoms or ions undergo decrease in oxidation state. The agent that causes reduction to occur is called the reducing agent and is itself oxidized in the process.

An oxidation number can be defined as the charge that an atom would have if both of the electrons in each bond were assigned to the more electronegative element. The term oxidation state is used interchangeably with the term oxidation number.

The following are the basic rules for assigning oxidation numbers:

1. The oxidation number of any element in its elemental form is zero.

2. The oxidation number of any simple ion (one atom) is equal to the charge on the ion.

3. The sum of all of the oxidation numbers of all of the atoms in a neutral compound is zero.

(More generally, the sum of the oxidation numbers of all of the atoms in a given species is equal to the net charge on that species.)

14.2 BALANCING OXIDATION–REDUCTION REACTIONS USING THE OXIDATION NUMBER METHOD

14.2.1 THE OXIDATION –NUMBER–CHANGE METHOD

1. Assign oxidation numbers to each atom in the equation.

2. Note which atoms change oxidation number, and calculate the number of electrons transferred, per atom, during oxidation and reduction.

3. When more than one atom of an element that changes oxidation number is present in a formula, calculate the number of electrons transferred per formula unit.

4. Make the number of electrons gained equal to the number lost.

5. Once the coefficients from step 4 have been obtained, the remainder of the equation is balanced by inspection, adding H^+ (in acid sol'n), OH^- (in basic sol'n), and H_2O, as required.

14.3 BALANCING REDOX EQUATIONS: THE ION–ELECTRON METHOD

14.3.1 THE ION–ELECTRON METHOD

1. Determine which of the substances present are involved in the oxidation-reduction.

2. Break the overall reaction into two half-reactions, one for the oxidation step and one for the reduction step.

3. Balance for mass (i.e., make sure there is the same number of each kind of atom on each side of the equation) for all species except H and O.

4. Add H^+ and H_2O as required (in acidic solutions), or

OH$^-$ and H$_2$O as required (in basic solutions) to balance O first, then H.

5. Balance these reactions electrically by adding electrons to either side so that the total electric charge is the same on the left and right sides.

6. Multiply the two balanced half-reactions by the appropriate factors so that the same number of electrons is transferred in each.

7. Add these half-reactions to obtain the balanced overall reaction. (The electrons should cancel from both sides of the final equation.)

14.4 NON-STANDARD-STATE CELL POTENTIALS

For a cell at concentrations and conditions other than standard, a potential can be calculated using the following Nernst equation:

$$E_{cell} = E^0_{cell} - \frac{.059}{n} \log Q$$

where E^0_{cell} is the standard-state cell voltage, n is the number of electrons exchanged in the equation for the reaction, and Q is the mass action quotient (which is similar in form to an equilibrium constant).

For the cell reaction

$$Zn + Cu^{2+} \rightarrow Cu + Zn^{2+}, \text{ the term } Q = [Zn^{2+}]/\{Cu^{2+}\}.$$

the Nernst equation takes the form:

$$E = E^0 - \frac{.059}{n} \log \frac{[Zn^{2+}]}{[Cu^{2+}]} .$$

14.5 ELECTROLYTIC CELLS

Reactions that do not occur spontaneously can be forced to take place by supplying energy with an external current. These reactions are called electrolytic reactions.

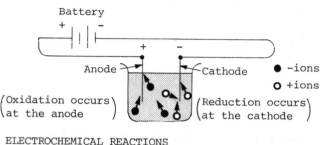

Battery

Anode — Oxidation occurs at the anode

Cathode — Reduction occurs at the cathode

● −ions
O +ions

ELECTROCHEMICAL REACTIONS
In electrolytic cells, electrical energy is converted into chemical energy.

14.6 FARADAY'S LAW

One faraday is one mole of electrons.

(1F = 1 mole of electrons; F is the "faraday".)

1 F ~ 96,500 coul; the charge on 1 F is approximately 96,500 coulombs.

One coulomb is the amount of charge that moves past any given point in a circuit when a current of 1 ampere (amp) is supplied for one second. (Alternatively, one ampere is equivalent to 1 colomb/second.)

Faraday's law states that during electrolysis, the passage of 1 faraday through the circuit brings about the oxidation of one equivalent weight of a substance at one electrode (anode) and reduction of one equivalent weight at the other electrode (cathode). Note that in all cells, oxidation occurs at the anode and reduction at the cathode.

14.7 VOLTAIC CELLS

One of the most common voltaic cells is the ordinary "dry cell" used in flashlights. It is shown in the drawing below, along with the reactions occurring during the cell's discharge:

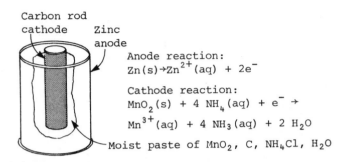

Anode reaction:
$$Zn(s) \rightarrow Zn^{2+}(aq) + 2e^-$$

Cathode reaction:
$$MnO_2(s) + 4\ NH_4^+(aq) + e^- \rightarrow Mn^{3+}(aq) + 4\ NH_3(aq) + 2\ H_2O$$

Moist paste of MnO_2, C, NH_4Cl, H_2O

In galvanic or voltaic cells, the chemical energy is converted into electrical energy.

In galvanic cells, the anode is negative and the cathode is positive (the opposite is true in electrolytic cells).

The force with which the electrons flow from the negative electrode to the positive electrode through an external wire is called the electromotive force, or emf, and is measured in volts (V):

$$1V = \frac{1J}{coul.}$$

The greater the tendency or potential of the two half-reactions to occur spontaneously, the greater will be the emf of the cell. The emf of the cell is also called the cell potential, E_{cell}. The cell potential for the Zn/Cu cell can be written

$$E^0_{cell} = E^0_{Cu} - E^0_{Zn} \qquad \text{where the } E^0\text{'s are standard reduction potentials.}$$

The overall standard cell potential is obtained by

subtracting the smaller reduction potential from the larger one. A positive emf corresponds to a negative ΔG and therefore to a spontaneous process.

$$\Delta G = -nFE$$

is the Nernst equation.

Also,

$$E = E^0 - \frac{RT}{nF} \ln Q$$

and

$$E = E^0 - \frac{.059}{n} \log Q$$

which is analogous to

$$\Delta G = \Delta G^0 + RT \ln Q.$$

THE PERIODIC TABLE

METALS

NONMETALS

KEY

112.40
Cd ← Symbol
48 ← Atomic number

Atomic weight → 112.40

TRANSITION METALS

PERIODS	IA	IIA	IIIB	IVB	VB	VIB	VIIB	VIII			IB	IIB	IIIA	IVA	VA	VIA	VIIA	O
1	1.0079 **H** 1																1.0079 **H** 1	4.00260 **He** 2
2	6.94 **Li** 3	9.01218 **Be** 4											10.81 **B** 5	12.011 **C** 6	14.0067 **N** 7	15.9994 **O** 8	18.9984 **F** 9	20.179 **Ne** 10
3	22.9898 **Na** 11	24.305 **Mg** 12											26.9815 **Al** 13	28.086 **Si** 14	30.9738 **P** 15	32.06 **S** 16	35.453 **Cl** 17	39.948 **Ar** 18
4	39.098 **K** 19	40.08 **Ca** 20	44.9559 **Sc** 21	47.90 **Ti** 22	50.9414 **V** 23	51.996 **Cr** 24	54.9380 **Mn** 25	55.847 **Fe** 26	58.9332 **Co** 27	58.71 **Ni** 28	63.546 **Cu** 29	65.38 **Zn** 30	69.72 **Ga** 31	72.59 **Ge** 32	74.9216 **As** 33	78.96 **Se** 34	79.904 **Br** 35	83.80 **Kr** 36
5	85.4678 **Rb** 37	87.62 **Sr** 38	88.9059 **Y** 39	91.22 **Zr** 40	92.9064 **Nb** 41	95.94 **Mo** 42	98.9062 **Tc** 43	101.07 **Ru** 44	102.9055 **Rh** 45	106.4 **Pd** 46	107.868 **Ag** 47	112.40 **Cd** 48	114.82 **In** 49	118.69 **Sn** 50	121.75 **Sb** 51	127.60 **Te** 52	126.9046 **I** 53	131.30 **Xe** 54
6	132.9054 **Cs** 55	137.34 **Ba** 56	57-71 *	178.49 **Hf** 72	180.9479 **Ta** 73	183.85 **W** 74	186.2 **Re** 75	190.2 **Os** 76	192.22 **Ir** 77	195.09 **Pt** 78	196.9665 **Au** 79	200.59 **Hg** 80	204.37 **Tl** 81	207.2 **Pb** 82	208.9804 **Bi** 83	(210) **Po** 84	(210) **At** 85	(222) **Rn** 86
7	(223) **Fr** 87	(226.0254) **Ra** 88	89-103 †	(260) **Ku** 104	(260) **Ha** 105													

* LANTHANIDE SERIES

138.9055 **La** 57	140.12 **Ce** 58	140.9077 **Pr** 59	144.24 **Nd** 60	(145) **Pm** 61	150.4 **Sm** 62	151.96 **Eu** 63	157.25 **Gd** 64	158.9254 **Tb** 65	162.50 **Dy** 66	164.9304 **Ho** 67	167.26 **Er** 68	168.9342 **Tm** 69	173.04 **Yb** 70	174.97 **Lu** 71

† ACTINIDE SERIES

(227) **Ac** 89	232.0381 **Th** 90	231.0359 **Pa** 91	238.029 **U** 92	237.0482 **Np** 93	(242) **Pu** 94	(243) **Am** 95	(245) **Cm** 96	(245) **Bk** 97	(248) **Cf** 98	(253) **Es** 99	(254) **Fm** 100	(256) **Md** 101	(253) **No** 102	(257) **Lr** 103